Forward

African American scientists and inventors have made and are making invaluable contributions to the scientific world. We are largely familiar with the accomplishments of pioneers like Dr. Charles Drew, Dr. Daniel Hale Williams, Benjamin Banneker, and George Washington Carver, but there are so many more scientists and inventors that shaped and are shaping the scientific world. This book identifies African American scientists and inventors that may be unknown to us, but not to those in the scientific community. It is divided into different scientific disciplines, with scientists' biographies and information on each scientists' contribution. Scientific experiments and methods are included to demonstrate and expand on some of the scientific principles discussed in the scientists' biographies. This book will serve as a great tool for science teachers and parents to show the contributions of African Americans in science that can be used during Black History month and beyond.

BIOLOGY

AND

LIFE SCIENCE

Emmett W. Chappelle

Emmett W. Chappelle was born in Phoenix, AZ in 1925. Mr. Chappelle received his BS degree in biochemistry from the University of California. He later received his master's degree in biochemistry from University of Washington. Mr. Chappelle taught at Meharry Medical College in Nashville, and he served as a research associate at Stanford University, Hazelton Laboratories, and NASA.

Mr. Chappelle's focus of research was in the area of bioluminescence, the production of light from a living organism. His major achievement in this area involved helping to develop methods to detect bacteria in water. This discovery was instrumental in improving the diagnoses of urinary tract infections. This technique is also used to detect bacteria in other fluids such as blood and drinking water, as well as some foods.

Ernest E. Just, PhD

Ernest E. Everett Just was born in 1883 in Charleston, SC. He received his BS degree in zoology, and he served as a professor at Howard University. He later became head of both the zoology and biology departments at Howard. Dr. Just was the first recipient of the Spingarn Medal awarded by the NAACP for outstanding achievement. He then earned a doctorate in zoology from the University of Chicago, where his main focus turned to studies in marine life and their mammal cells.

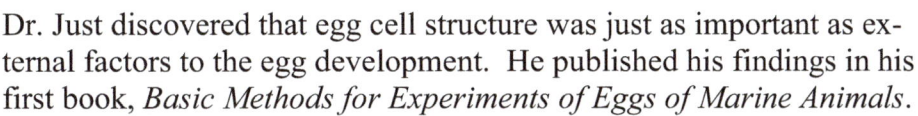

Dr. Just discovered that egg cell structure was just as important as external factors to the egg development. He published his findings in his first book, *Basic Methods for Experiments of Eggs of Marine Animals*. After finishing his book, Dr. Just moved to Europe to escape the rampant racism and discrimination in the US. While in Europe, he published his most famous work, *The Biology of the Cell Surface*. His study of the cell involved the transport of water, and the effects of light on chromosomes in the cell. He eventually returned to the US, after the beginning of World War II.

Cell Transport

Dr. Just studies involved cells and the movement of materials across them. We all know that cells make up all living things. In this investigation, we will see how cells are affected by substances that are transported into and out of them.

Substances are transported in and out of our cells through three main methods across the cell membrane: diffusion, osmosis, and active transport.

*Which of these three methods require energy? _____

For this investigation, we will look at diffusion and osmosis.

Diffusion is the movement of molecules from an area of high concentration to an area of low concentration, until **equilibrium** is achieved. Equilibrium is achieved when there is the same amount of molecules on each side of the membrane. See the diagram below.

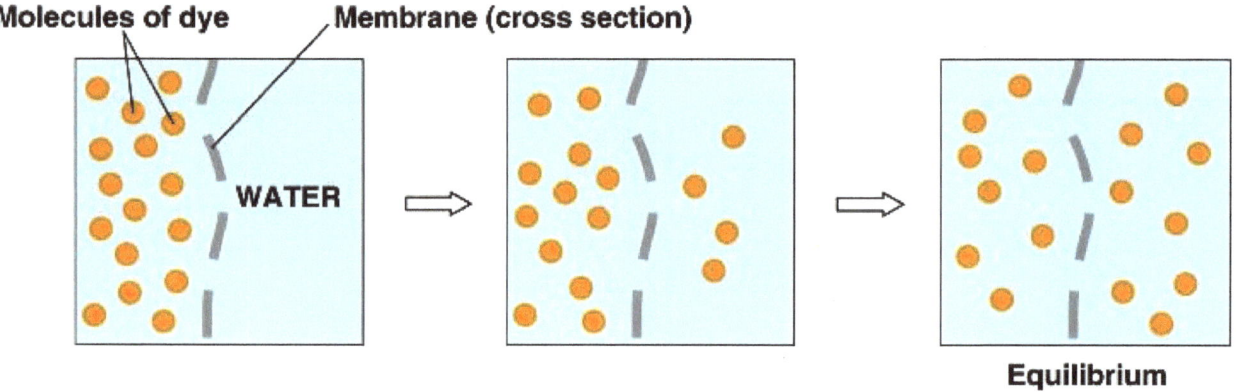

Diffusion is important in allowing cells to maintain a healthy balance of nutrients and other materials that they may encounter!

Osmosis is the same as diffusion, but it describes the movement of water only.

In order to carry out this investigation, you will need the following items:

MATERIALS
1 egg (more if you are not careful!)
A plastic bowl or container
Vinegar
Water
Food coloring
Salt
A liquid of your choice (soda, juice, liquid detergent, etc.)
Ruler
String

PROCEDURE

Egg preparation

1. Fill the plastic container with vinegar and place an egg in the container, making sure that the egg is completely covered with vinegar. Let the egg stay in the vinegar for at least 24 hours. The shell should separate away from the egg. If the shell does not fall off the egg easily, it may need to stay in the vinegar a little longer.

2. When all of the shell comes off of the egg, you will have an intact, but fragile, egg. Clean the plastic container, and after carefully rinsing the egg, place it in the clean container.

3. Take the string and carefully wrap it around the middle of the egg. Holding the string where the two ends meet, use the ruler to measure the length of the string. Record this measurement here

Initial circumference of the egg   _____(cm)

Water

4. Place the egg in the plastic container. Pour in enough water to completely cover the egg. Let the egg stay in the water for 2 days.

5. After 2 days, carefully take the egg out of the water, leaving the water in the container.

6. Measure and record the circumference of the egg, using the method described in step 3.

Circumference of the egg, after 2 days in water   _____(cm)

Compare this measurement with the initial measurement from step 3.

_____

_____

_____

Food coloring

7. Place the egg back into the water, and drop in 7-10 drops of food coloring.

8. After 2 days, carefully take the egg out of the colored water, and pour the colored water out of the container. Clean the container, and set aside.

Describe the appearance of the egg, measure and record the circumference of the egg, using the method in step 3.

Appearance of the egg:

_____

_____

Circumference of the egg in colored water _____(cm)

Salt water

9. Fill the container again with water, and place the egg back in the water, add a good amount of salt. Place the egg in the salt water mixture for about 2 days.

10. After 2 days, carefully take the egg out of the salt water mixture, and pour the mixture out of the container. Measure and record the circumference of the egg, using the method in step 3.

Circumference of the egg in salt water _____(cm)

Compare this measurement to measurements taken in steps 3 and 10.

_____

_____

_____

_____

_____

11. Fill the cleaned container with the liquid of your choice. Place the egg in the container with your chosen liquid for about 2 days. Predict what you think will happen to the circumference of the egg after 2 days.

_____

_____

_____

_____

12. After 2 days, carefully take the egg out of your liquid. Pour the liquid out of the container, clean, and put away. Measure the circumference of the egg, and record here.

Circumference of the egg in your liquid _____**(cm)**

Compare this measurement to all of the other measurements taken. In your comparison, include whether or not your prediction was correct.

_____

_____

_____

_____

_____

_____

_____

_____

_____

_____

Questions (Answer using complete sentences)

How was the method of osmosis demonstrated in this investigation?

_____

_____

_____

Explain why there was a change in the circumference of the egg in step #3 (initial) to the circumference of the egg in step #6 (after 2 days in water).

_____

_____

_____

What happened to circumference of the egg after it was in the salt water for 2 days?

_____

_____

Why? (Think of the process of osmosis)

_____

_____

_____

What do you think would happen if you placed the egg in water from the beach? Why?

_____

_____

_____

_____

_____

Using what you know about the process of diffusion, explain the changes of the egg when it was placed in water with food coloring.

_____

_____

_____

_____

_____

_____

_____

**Make a bar graph that compares the circumference of the egg with each liquid it was placed in for 2 days. Your graph should have 5 bars. Be sure to label each axis and give your graph a title.

**See the example below. If your computer has Microsoft Excel, see how you can use it to generate your own graph. Be sure to label it!

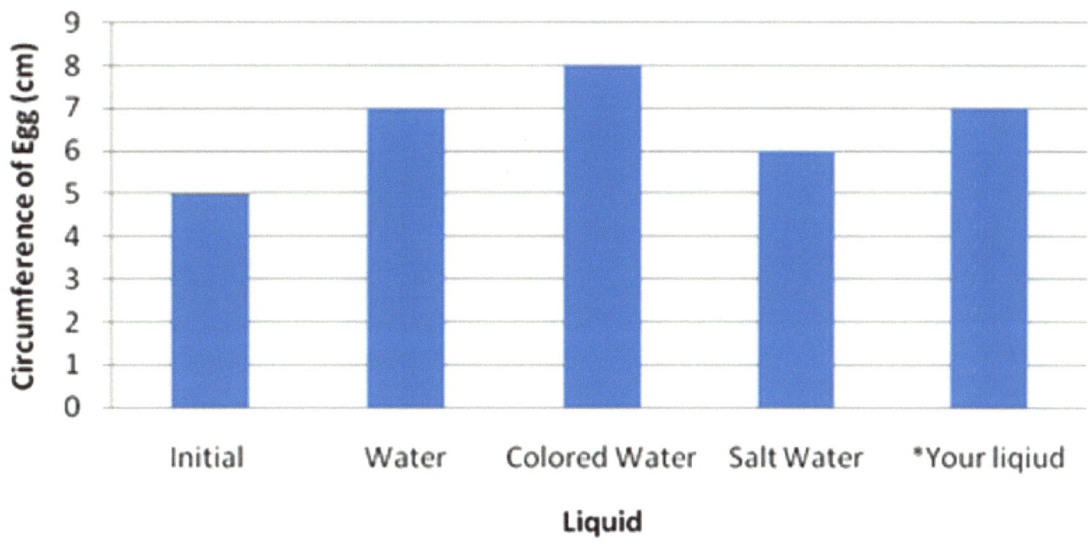

PHYSICS

Shirley Ann Jackson, PhD

Shirley Ann Jackson was born in 1946 in Washington, DC. She was always fascinated by science and she cultivated this interest by attending advanced classes in science and math throughout high school. Dr. Jackson attended MIT for her bachelor's and doctorate degrees in the area of physics with a concentration in elementary particle theory. She was very interested in studying subatomic particles found in the atom. This led her to conduct research at several prestigious laboratories in the US and Europe.

In 1976 Dr. Jackson joined AT&T Bell Labs where she made significant progress in the exploration of the theory regarding particle charge density and the behavior of electrons in different surfaces. She has received several honors, including the Thomas Alva Edison science award and the CIBA-GEIGY exceptional black scientist award.

In 1995 Dr. Jackson, was appointed head of the Nuclear Regulatory Commission by Pres. Bill Clinton. In 1999, Ms. Jackson was named president of Rensselaer Polytechnic Institute, and she holds 45 honorary degrees. She was inducted into the National Women's Hall of Fame in 1998, and named a fellow of the Association for Women in Science, thus bolstering her desire to actively promote women in science.

Herman Branson, PhD

Herman Branson was born in Pocahontas, VA in 1914. He earned his BS degree from Virginia State College, and his PhD in physics from the University of Cincinnati. Dr. Branson's research and interests were in the fields of mathematical biology and protein structure. He was the co-inventor of the alpha helix, which was arguably one of the most important discoveries in the last century. His mathematical model and analysis of the alpha helix is considered a necessary foundation for protein structure and behavior. Dr. Branson was not mentioned as a contributor to the finding, which excluded him from recognition as his colleagues were awarded the Nobel Prize for the discovery.

Despite this major snub, Dr. Branson continued his research on the connection between physics and biology, studying topics such as radioactive and stable isotopes in biological transport. He also conducted physical and chemical research on sickled red blood cells, which causes disorders that affect many African Americans. Dr. Branson served as the president of Central University in Wilberforce, OH from 1968-1971. He then became the president of Lincoln University in Pennsylvania, and served until he retired in 1985.

11

Atom Activities

Dr. Jackson's interest in particle theory began with developing an understanding of atomic structure. These activities will help you to visualize the atom, and what makes it work!

Atoms and Element Structure

The particles that make up atoms have either a positive charge, a negative charge, or no charge at all! These particles are called **subatomic** particles. Find the names of those particles and their charges, and write them here.

1. _____

2. _____

3._____

Items in the world are made up of elements. An element is made up of many of the same atoms. The atoms are arranged with a nucleus and energy levels.

Which subatomic particle(s) are located in the nucleus?

Which subatomic particle(s) are located in the energy levels?

How many of those subatomic particles can fit in each energy level?

Energy level 1 –

Energy level 2 –

Energy level 3 –

Atom models

Here is a list of three elements that you should be familiar with. Look up the number of subatomic particles and energy levels in each element. Write the answers next to the appropriate element.

Example: Boron (B): Protons—5, Neutrons—5, Electrons—5, Energy levels—2

Helium (He)

Carbon (C)

Oxygen (O)

Making the model

A. Assign color to different subatomic particles using, gumdrops, M&Ms, etc.

B. Using a large piece of construction paper or poster board, draw an atom with a nucleus, and the corresponding energy levels for the elements listed above. Be sure to determine how many particles are allowed on each energy level, and which particles are located in the nucleus.

C. For each element, place the subatomic particles in your atom drawing in the correct places.
At the bottom, or next to your model, make a key that explains what item is used in the place of the subatomic particle.

Example: The element is Boron (B)

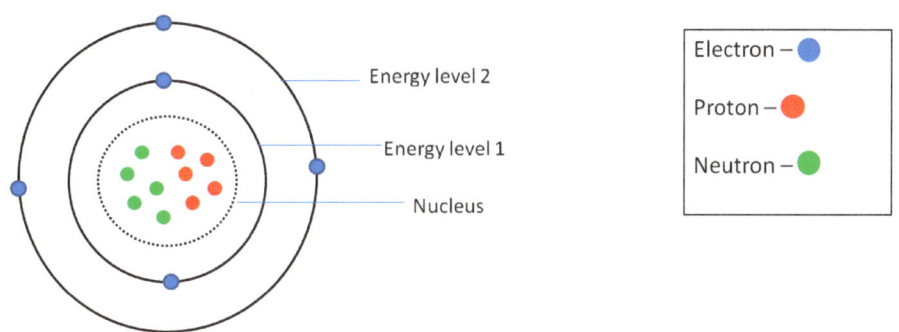

CHEMISTRY

Augustus Jackson

The father of modern Ice Cream

Augustus Jackson was a former White House chef for President Andrew Jackson. He is credited for creating the modern method of manufacturing ice cream. His methods employed the science of lowering the freezing point of a solution (freezing point depression) by using salt. Mr. Jackson moved from Washington, DC to Philadelphia, PA where he developed many popular ice cream flavors which were distributed to ice cream parlors all over Philadelphia. His tin packed ice cream was thought to be the best in the city. Mr. Jackson is also sometimes referred to as the father of ice cream.

Percy Lavon Julian, PhD

Chemistry is a science that explores the composition of matter in terms of elements and compounds. Understanding how elements and compounds react with each other gives insight and explanation of natural processes. Synthetic chemistry is using simple compounds to design and build more complex compounds. Synthetic chemistry is important to developing compounds that can be used in such disciplines as food chemistry and medicine.

Percy Lavon Julian was a hugely successful synthetic chemist who developed products that are widely used today. Dr. Julian was born in Montgomery, AL in 1899. He received his BA degree from DePauw University. He went on to earn a MA from Harvard University, and his PhD from the University of Vienna. After teaching stints at West Virginia State University and Howard University, Dr. Julian accepted a position as a research fellow at his alma mater, DePauw University. During his time at DePauw, he developed the total synthesis of physostigmine, a drug used in the treatment of glaucoma. This new drug was considerably more affordable than current treatments at that time. His discovery made him the most sought after synthetic chemist at that time. He then took a position as a research chemist at Glidden Company in Chicago, where he developed products using soybeans, including the female hormone progesterone, a hormone essential to helping women avoid miscarriages. He was so successful at Glidden that he decided to start his own company, Julian Laboratories. At his laboratory Dr. Julian continued his research on soybeans. He then discovered a way to synthesize cortisone using soybeans. This discovery was major as his method only cost about 50 cents per gram, as opposed to the animal derived cortisone which cost about $700 per gram! This helped those who suffered from conditions such as arthritis, to have the ability to afford the medicine to treat their ailments. Dr. Julian employed many African American scientists in his laboratory, and he eventually sold it 1961 to Smith, Kline, and French, the predecessor of GlaxoSmithKline. He was also the first African American chemist to be elected to the National Academy of Sciences in 1973.

15

The Science Behind Ice cream

The making of ice cream is a great way to demonstrate the science of freezing point depression. Freezing point depression is achieved when the freezing point of a substance is lowered using another substance. In the case of making ice cream, the freezing point of the ice used to freeze the ice cream mixture is lowered using salt. Since larger salt crystals take longer to dissolve, which keeps the ice cooler longer, rock salt is used in ice cream making as opposed to table salt. The following activity of making ice cream is a great way to demonstrate freezing point depression, with a treat! It can get messy, so be sure to use lab aprons, and thick gloves!

I scream, you scream, we all scream for ice cream!

Materials
1 cup of whole milk (240 mL)
2 tablespoons of sugar (brown or white)(28.3 g)
½ teaspoon of vanilla extract
Crushed ice
Rock salt
One quart size reclosable bag
One gallon size reclosable bag
1 plastic spoon
1 cup
1 apron
2 heavy gloves
*optional ingredients: chocolate syrup, fruit, or whatever you like in your ice cream!

Procedure

1. Fill the quart sized reclosable bag with the milk, sugar, vanilla extract, and optional ingredients like chocolate syrup, pureed fruit, etc.
2. SECURELY close the bag!
3. Fill the gallon sized reclosable bag HALFWAY with ice.
4. Pour about 45 mL of rock salt into the gallon sized bag with ice.
5. Place the quart sized bag into the gallon sized bag and close the gallon bag SECURELY.
6. Put on the gloves and shake, flip, or massage the quart sized bag around, until the mixture freezes. This will take approximately 10-15 minutes.
7. When the ice cream is solid, remove the ice cream bag from the ice bag. Discard the ice bag in the trash, and spoon out the ice cream into a cup.
Top the ice cream with you favorite toppings, and enjoy!

MATHEMATICS
AND
ENGINEERING

J. Ernest Wilkins, Jr., PhD

Jesse Ernest Wilkins, Jr. was born in 1923 in Chicago, Illinois. His mother was an accomplished schoolteacher, and his father was a successful attorney, so the importance of education was instilled in him early in his life. He entered his parents' alma mater, the University of Chicago, at age 13. Dr. Wilkins received his BS degree in mathematics at age 16, and went on the complete his MS and PhD degrees at age 17 and 19, respectively. He started his career as a professor of mathematics at Tuskegee Institute (University), and went on to work as a mathematician at several laboratories and companies, and even earned his BS and MS degrees in mechanical engineering from NYU. Dr. Wilkins even contributed to the Manhattan Project, the project undertaken during World War II that developed the first atomic bomb. As a professor of mathematics at Howard University, Dr. Wilkins helped to start Howard's Ph.D. program in mathematics, the first at a historically black college. One of his most noted contributions to his field of study was his development of a mathematical model to explain gamma radiation and how to shield against it. Dr. Wilkins has published several papers and has received many awards from national mathematical and scientific organizations, as well as the United States military. He came out of retirement in 1990, and accepted a position as Distinguished Professor of Applied Mathematics and Mathematical Physics at Clark Atlanta University. Dr. Wilkins is considered one of the greatest mathematical minds of this time.

Archibald Alphonso Alexander

Archibald Alexander was born in Iowa in 1888. He was a gifted athlete, who worked and attended school throughout high school and college. He attended Highland Park College, and entered the college of engineering at State University of Iowa (Iowa State University) in 1909. Alexander played football, and Iowa State was one of the few colleges that would allow blacks to play at that time. He went on to graduate with a civil engineering degree in 1912. He worked as a design engineer with a company that built bridges, and after only 2 years, he started his own firm. He traveled to London, England to study bridge design, and upon his return garnered several large building contracts with his alma mater, Iowa State University. During World War II, Alexander's firm was commissioned to construct an airfield in Tuskegee, AL, which served as the training facility for the black Air Force battalion known as the Tuskegee Airmen. His firm branched out into Washington D.C. where they built the fames Tidal Basin Bridge, as well as other bridges and highways. His work even went international with several projects completed in Venezuela and Puerto Rico. He was later nominated for governor of the US Virgin Islands by President Eisenhower. He only served 16 months, and died shortly after.

BRIDGE BUILDING

Archibald Alexander started out as a civil engineer, interested in bridge design and building. Bridges are made out of a variety of materials depending on their location and function. A bridge with a low degree of traffic is sometimes built differently from a bridge that has a high degree of traffic. The strength of the bridge is essential to its function. How many times have you seen stories about a bridge collapsing? How much pressure can a bridge take, and what are the best materials to use?

In this investigation you will use several different materials to build bridges. Your bridge can be any design that you believe will hold the most weight. This may require some research on your part. You would want to research bridge design. After you build the bridges, you will test their strength using weighted materials like books!

Materials
(Required)
Toothpicks
Drinking straws
Spaghetti
(Optional)
Marshmallows
String
Gumballs
Spice drops
Glue

MEDICINE

Patricia Bath, MD

Patricia Era Bath was born in Harlem, New York in 1942. Her love of science led her to many opportunities, from the editor of her school's science paper to her selection as a cancer research assistant at the age of 16. She earned her Bachelor's degree from Hunter College, and her MD from Howard University. After graduating with honors, Dr. Bath began a fellowship in the area of ophthalmology at Columbia University. While working at the Harlem Eye clinic, she witnessed the disparities in eye health between blacks and whites. She decided to combat this problem by initiating Community Ophthalmology. This initiative provided screening for general vision, glaucoma, and cataracts to underserved communities. She completed her residency in ophthalmology, and became a faculty member at UCLA. In her research at UCLA, Dr. Bath developed the "Laserphaco Probe". This device would remove cataracts using lasers. The device made cataract removal process less painful and more accurate. It also helped to restore sight to some patients that were previously blinded by cataracts. She had the device patented, and it is used for eye surgery all over the world.

Samuel Lee Kountz, Jr., MD

Samuel Kountz, Jr. was born in Lexa, Arkansas in 1930. His father was a Baptist minister, who also served as a nurse for the sick in the small town. Watching his father care for the sick inspired Samuel Kuntz to want to become a doctor. Although he failed at his first attempt at college, he never gave up. He eventually was accepted into the University of Arkansas Medical School, along with the first African American students to attend the school. After graduating from medical school, Kountz entered a surgical residency program where he began his study in organ transplantation. During his residency, he assisted in the first kidney transplant between a donor and a recipient that were not identical twins. He decided to further research kidney transplantation. His research helped to identify a hormone that greatly decreased organ rejection which allowed for greater transplant success. This breakthrough was significant as it increased the time a transplanted organ could successfully function in a new body. He also helped to develop a device that increased the preservation of donated kidneys. Throughout his career, he performed over 500 transplants, including the first one in Egypt in 1965.

O, Say Can You See? Better!

Dr. Bath's dedication to making sure everyone has the ability to see has made a major impact on people's lives today. This investigation explores the affects of cataracts on vision, and how they can affect your life.

Study Activity – Ophthalmology

Cataracts

What are cataracts?

How do they form?

How do they affect sight?

How are they treated?

Diagrams (labels)

Cataracts

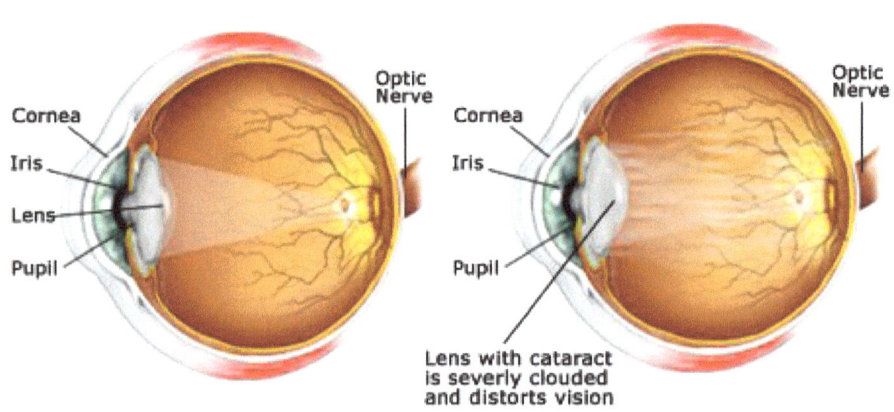

Difference in sight activity

Stand about 3 feet from the picture above. Are you able to clearly read the labels on the diagram of the eye? On a scale from 1 to 10, with 10 being perfectly clear, how clear are the diagram labels?

Now hold a clear plastic bag about 3 inches from your eye, and try to read the diagram labels. On a scale from 1 to 10, with 10 being perfectly clear, how clear are the diagram labels? Can you see a difference?

This is a very simple way to see how cataracts affect your ability to clearly see and distinguish items in front of you.

Answer the following questions.

1. What are some things that are safer to do when you can see clearer?

2. What are some things that are hard to do without seeing clearly?

Do some research on cataracts in your local library. Find an ophthalmologist in your area, and interview them on the formation, prevalence, treatment and prevention of cataracts. Use all of the information to develop a short educational pamphlet to educate people about cataracts.

(Pamphlet templates can be found using computer office programs like MS Word, or by finding other templates online! Be sure to include pictures or eye catching graphics to make you pamphlet stand out! You may even present it to eye clinics to aid in their patient's education on cataracts.)

REFERENCES

Simmons, Vivian O., Blacks in Science and Education. Washington, D.C.: Hemisphere Publishers, 1989

Whitmore, Todd. , www.chemistryexplained.com

Houston, Johnny L., NAM Newsletter, Fall Issue, 1994

Gray, Madison, "Black History Unsung Heroes", www.time.com, January 12, 2007

Brown, Mitchell. https://webfiles.uci.edu/mcbrown/display/faces.html, 1995-2000.

Stradley, Linda. "Legends and Myths of Ices and Ice Cream", www.whatscookingamerica.net

Williams, Dr. Scott. "Physicists of the African Diaspora"and "Mathematicians of the African Diaspora", http://www.math.buffalo.edu/mad/physics/physics-peeps.html

Davidson, Martha. "Changing the Face of Medicine: Celebrating America's Women Physicians", http://www.nlm.nih.gov/changingthefaceofmedicine/physicians/biography_26.html

http://www.chemheritage.org/discover/online-resources/chemistry-in-history/themes/

PHOTO CREDITS

"Emmett Chappelle". Photo. Courtesy of NASA. invent.org.2007. 2 February 2012 <www.invent.org>.

"Ernest E Just". Photo. musc.edu. 5 January 2010. <www.musc.edu>

"Shirley Ann Jackson". Photo. rpi.edu 10 January 2010. <www.rpi.edu>

"Herman Branson". Photo. Lincoln University, Special Collections and Archives. pnas.org. 1 January 2010. <www.pnas.org>

"Percy Lavon Julian". Photo. Gift of Ray Dawson, CHF Collections. chemheritage.org. 10 January 2010. <www.chemheritage.org>

"J. Ernest Wilkins". Dan Dry, photographer. Photo. news-uchicago.edu. 19 February 2012. <news-uchicago.edu/releases>

"Patricia Bath". Photo. ossc.org. 19 February 2012. <www.ossc.org>

"Samuel Lee Kountz, Jr". Photo. downtownstate.edu. 6 May 2012. <www.downtownstate.edu>

"Archie Alexander". Charles Henry Alston, artist. 1943. Drawing. commons.wikimedia.org/wiki/File:Archie_alexander.jpg. 22 February 2013. <commons.wikimedia.org>

www.ingramcontent.com/pod-product-compliance
Lightning Source LLC
Chambersburg PA
CBHW041310180526
45172CB00003B/1038